# WHAT IF I LEARN™
# PHYSICS

**WRITTEN BY DAVID LIMPUS**
**ILLUSTRATED BY HAILEY LIMPUS**

donated to
Arlington Elementary School
by
**WHAT IF I LEARN™ MEDIA**
whatifilearn.com

JUNE 2023

Text Copyright © 2020 by David Limpus. All rights reserved.
Illustrations Copyright © 2020 by Hailey Limpus. All rights reserved.

Illustrations were created using Illustrator, Pages, and Painter.

Published by What If I Learn™ Media. No parts of this work may be reproduced, stored in a retrieval system, or distributed, in any form or by any means, including but not limited to, electronic, mechanical, photocopying, recording, and scanning, without written permission from the publisher.

What If I Learn™ Media
101 Creekside Crossing, Suite 1700-123
Brentwood, TN 37027

Printed in the United States of America
First Edition, April 2020

Library of Congress Control Number: 2019916480

ISBN 978-1-7331629-7-5 Hard Cover Edition
ISBN 978-1-7331629-8-2 E-Book Edition

whatifilearn.com

learning sparks innovation™

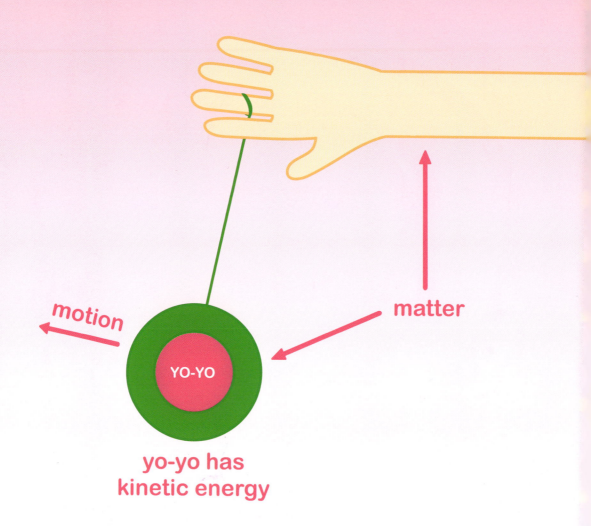

**Physics** explains how the universe works. **Physics** is the study of matter, energy, and motion.

# carbon atom

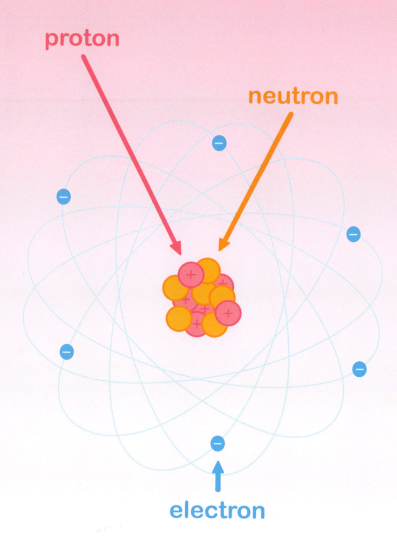

Atoms are the building blocks of matter. An atom contains subatomic particles called **electrons, neutrons,** and **protons. Electrons** are negatively charged. **Neutrons** have no charge. **Protons** are positively charged. **Neutrons** and **protons** form the nucleus, and **electrons** orbit the nucleus.

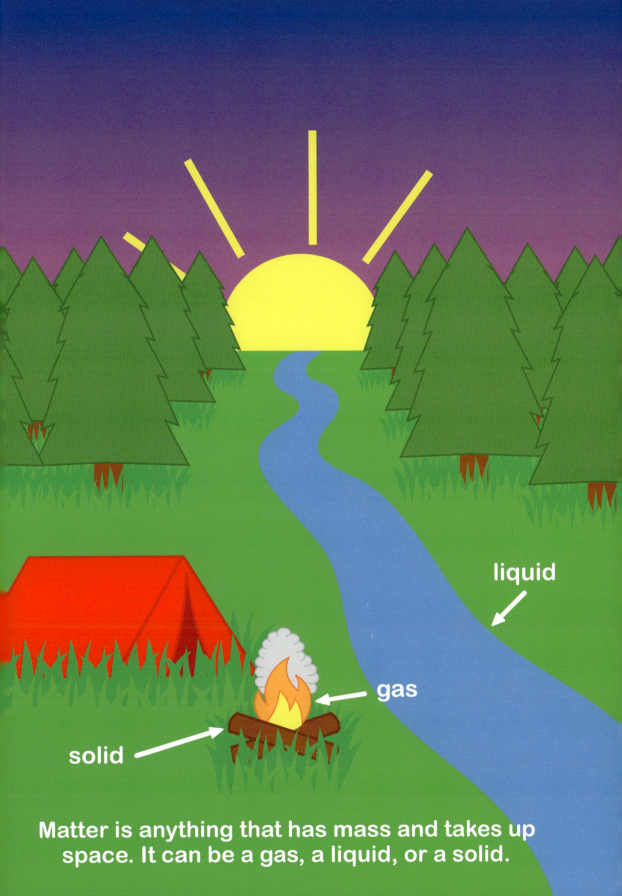

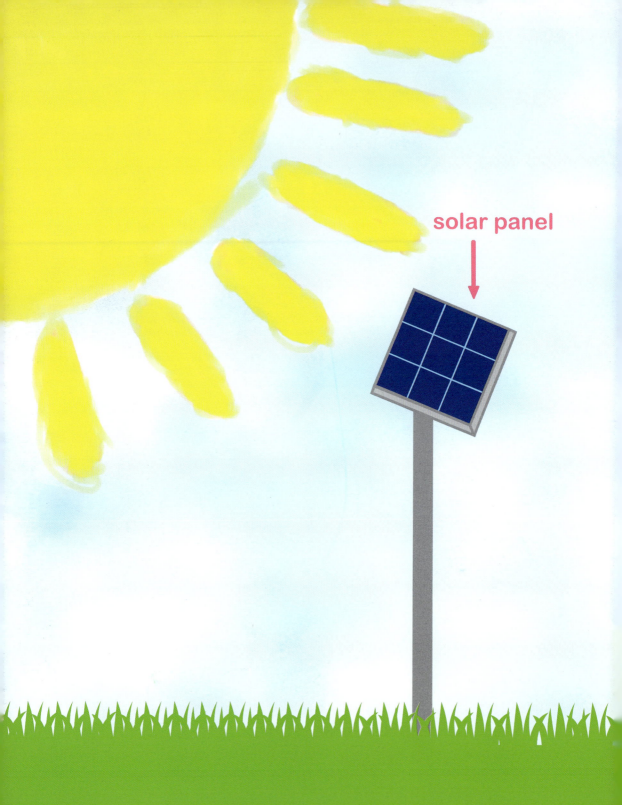

Energy is the ability to do work. Work is the transfer of energy from one object to another. The solar panel turns light energy from the sun into electrical energy.

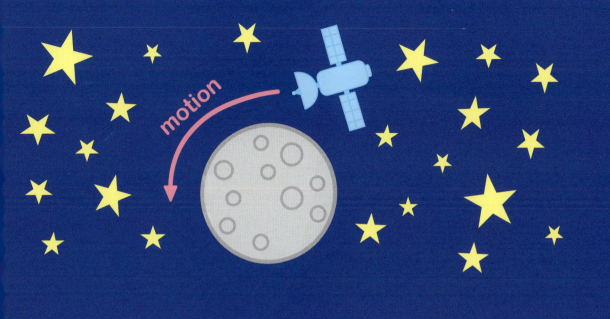

Motion is an object changing its position over time. An object in motion has kinetic energy. Motion happens when a force, such as a push or pull, acts on an object. The ball and the satellite have kinetic energy.

**Sir Isaac Newton**

**Sir Isaac Newton** mathematically explained motion and gravity during the 1600s. To this day, we use Newton's three laws of motion, and explanation of gravity.

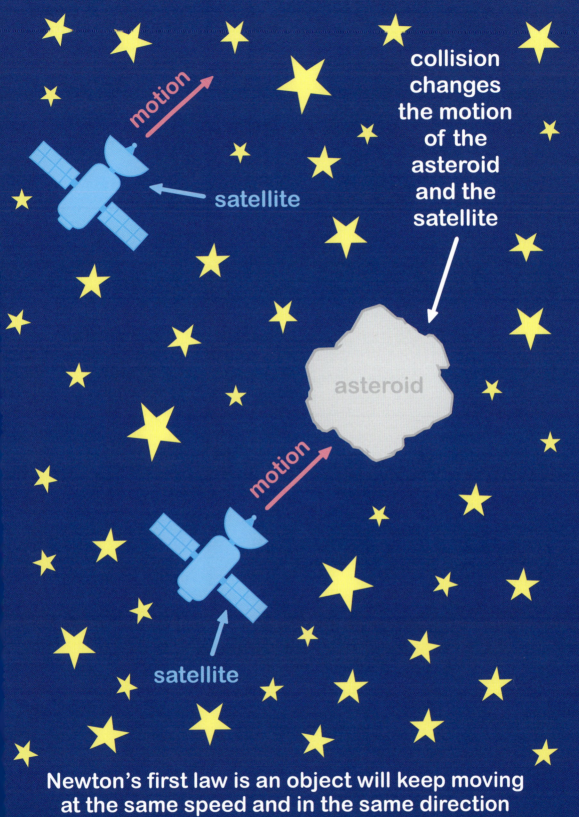

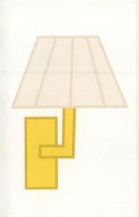
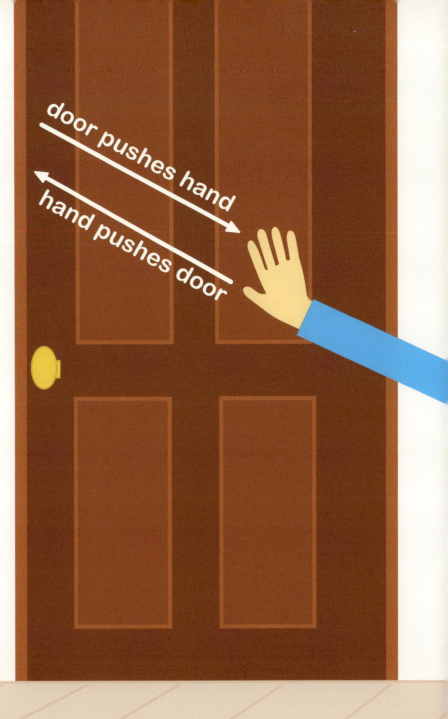

Newton's third law is every force has an opposite and equal reaction force. When you push on a door, the door pushes back on your hand with the same amount of force. This is how rockets launch into space!

Gravity is the force that pulls objects together. Anything that has mass pulls on other objects. The larger the mass, the stronger the pull. Earth's gravity keeps you from floating into space!

Albert Einstein

**Albert Einstein** was a **physicist**. One of his most famous equations is the mass-energy equivalence equation, $E=MC^2$.

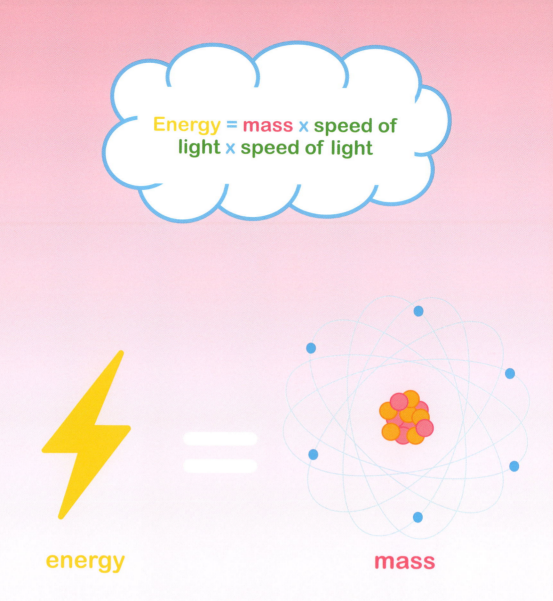

E=MC² is part of Albert Einstein's theory of special relativity. It means energy is equal to the mass of an object, multiplied by the speed of light, multiplied by the speed of light. The equation explains how energy and mass are the same thing, and how much energy is inside mass.